The Alma Project
By Simenona Martinez

The Alma Project is a neurological trauma treatment for treating complex post-traumatic stress syndrome in patients with Chronic Medical, Mental Health conditions and with Veterans Trauma Recovery.

This program is intended for treating veterans, refugees, and victims of violent, war or sex crimes.

The Alma Project

**Veteran's Recovery, KLS, ASD, Chronic Medical and Mental Health
Intensive Trauma Treatment**

The Almadate Center is a research and care facility for Veterans and other victims of C-PTSD related trauma.

The Environment:

The patients are housed in an inpatient facility. The key feature of the Alma Process is that all clocks are prohibited. The patients undergo a series of hypnosis sessions to induce trigger responses to work through the patient's processing and coping methodology while under doctors' care with medication management.

The Objective:

The objective is to overcome triggers associated with recall trauma and those associated with chronic illnesses.

The Patient Intake:

The intake process will record the patient's trauma inventory of all the shapes, colors, sounds, scents, textures, and the specific details associated with patient's trigger data.

The patient will also be interviewed regarding de-escalation data.

The Process:

The data is then classified and categorized in the Almadater Algorithm which formulates and computes this data into a simulation within bed pod. The patient is then placed into a pod with a bed that artificially reproduces these triggers.

This process is called:

Quantum Sequence Triplication and Quantum Operation Supremacy with Artificial Intelligence Methodology

Overview

Quantum Sequence Triplication and Quantum Operation Supremacy with Artificial Intelligence Methodology is a quantitative strategic system which trains A.I to have accurate rapid output data using Quantum Sequence Triplication and Quantum Operation Supremacy Methodology.

It analyzes various sequences from a master database which includes footage, information data, research information data, game theory ideology and Nash Equilibrium with database for optimum strategical outcomes.

A qubit or quantum bit is a basic storage/symbol, in which a bit of quantum information is stored/encoded-the quantum version of the classic binary bit physically realized with a two-state device. In Quantum Sequence Triplication, we use 3 qubit bits to achieve quantum supremacy without grey area bias. Qubit 1 and 2, are encoded using the Quantum Operation Supremacy methodology, with the exception of the data being opposing information and data. Finally, qubit 3, is the output of the data calculation process of opposing outcomes to determine the sum data devoid of grey area but most importantly, fact checked.

This is Quantum Supremacy.

The technology is intended for military strategy and defense, medicine and medical development without error, and the groundwork for quantum supremacy.

Quantum Operation Supremacy Overview: The use of traditional binary single qubit output methodology is the baseline for this operation and in the alteration regarding quantum operation. The system works by implementing artificial qubits to source and produce calculation data for equation execution. The data is stored and categorized into more ridged data columns to eliminate error and to achieve simplification. Sorting data by patterned repetitions in the level of importance relevance for desired outcome. This rapid data shuffling and detecting repetition sources regarding patterns can assist solving complex cryptology. Artificial qubits have infinite potential for quantum calculations. The use of frequency detection can detect slight use, alteration, or manipulation within the quantum monitoring system. This is an advancement towards code breaking within algorithmic systems such a reversal of Shor's algorithm, as well as CAPTCHA generated processes.

The De-escalation:

Upon completion of the trigger response session, the staff will then use the same process and procedures for the de-escalation regime.

The Cycles:

Length: 7-day Cycles.
Start: 4.20.21

<u>Patient A</u> entered the Almadate on 4/20/21.

The patient's KLS episode has begun, and she begins to show distress.

Kleine–Levin syndrome (KLS) is a rare sleep disorder mainly affecting teenage boys in which the main features are intermittent hypersomnolence, behavioral and cognitive disturbances, hyperphagia, and in some cases hypersexuality - Cleveland Clinic

She is afraid, disoriented, and unable to walk. She is also experiencing sensory overload due to the C-PTSD flare up, which has exacerbated her other underlying conditions such as Lupus, KLS, SPD, and ASD.

Night 1:

Medication, both western and eastern, can be used during this process with close physician and staff bedside observation.

- **The Trigger Data:** Establishing colors, sounds, textures, smells, number, and other associated trigger data. The understanding of fear and translating that experience into data for stimulation, treatment plan, and aftercare.
- **The De-escalation Data:** Determining the positive triggers which can be translated into data, to be used upon the completion of the Trigger Data Process.

Night 2:

The Pod Sensation Trauma Re-creation.

- The patient will experience sensory sensation triggers during this process, with the purpose of inducing the trauma response experienced during onset trauma recall seen in both C-PSTD and PTSD.
- Finally, the de-escalation process.

   (i.e., Sexual Assault, Desert Storm, Line of Fire injury, Medical Crisis, or Episode)

Night 3:

The Hypnosis with Pod Sensation Trauma Re-creation

This process is intended to trigger the trauma response to help rewrite the patient's response methodology.

Patient A:
- Induced meditative state.
- Awakened to triggering sensations, repeatedly in intervals, prompted by the classified trigger data input and overseen my staff.
- De-escalation process.

This process is essential in both KLS and C-PTSD because it allows the patient to relive these experiences with a sense of control while in a controlled environment with an incredibly intensive methodology to achieve better processing outcomes.

The de-escalation process is essential in this respect.

Night 4:

REPITITION: Hypnosis with Pod Sensation Trauma Re-creation.

This process is essential to avoid creating additional trauma while treating the prior.

Patient A:
- The patient repeats this process nightly.
- The de-escalation process begins its interval sequencing to ensure both the current and prior dream trauma is improving.

Night 6-7:

Hypnosis with Pod Sensation Trauma Re-creation and Sleep Soothing Therapy.

This process is intended to trigger the trauma response to help rewrite the patient's *essential* response methodology. In addition to creating the aftercare routine for trauma therapy maintenance.

Patient A:
- Induced meditative state.
- Awakened to triggering sensations, repeatedly in intervals, prompted by the classified trigger data input and overseen my staff.
- The patient sleeps with their pre-selected sleep white noise sound which is accompanied by a weighted blanket.

The process is intended to be repeated over the course of a cycle stay.

It must be accompanied by both traditional psychotherapy and medication management. The length of the patient's stay is dependent on the severity of the diagnoses.

This process requires hospitalization.

Aftercare is to be heavily monitored and observed.

In the case of KLS, during the trigger data inventory intake, 5 major featured objects are to be identified as sources of trauma and de-escalation patterns.

During the cycle, an item is to be taken away and replaced until the patient is fully awakened with one remaining item of de-escalation representation.

It is essential that upon waking, the patient identifies themselves in the mirror by name and where they are.

In cases outside of KLS, the patient can remove these items themselves, which can be incredibly therapeutic. This is process is called "Disarming the threat."

(i.e., Sexual Assault, Desert Storm, Line of Fire injury, Medical Crisis, or Episode)

During both episodes of KLS and C-PSTD with psychosis, patients can become frightened by items and people, with the idea that they are evil or even possessed.

This creates an associated trauma with those colors, objects, people, and sensory input. The association of trauma to objects is a present in ASD.

Treating the Subconscious:

Traumas within one's subconscious can manifest in the following:

- Nightmares
- Childhood Recall
- Sensory Response
- Music and Television

This patient's cycle ended on 4/27/21.

The patient engaged in art and music therapy until discharge.

More on aftercare, the practice of disarming threats can be repeated at home by mediation practices and methodology patient procedure learned during this process, specifically, being aware of one's <u>current</u> surroundings, listening for your soothing sound and the utilization of a weighted blanket.

Most importantly, a strong personal and medical support system in place upon discharge.

This treatment is used for the following:

- Veteran Recovery
- C-PTSD
- PTSD
- Autism Spectrum Disorder
- Narcissistic Abuse Recovery
- Kidnapping
- Childhood Trauma
- Cyberstalking
- Bullying
- Domestic Violence
- Sex Trafficking

The Alma Project
By Simenona Martinez

www.ingramcontent.com/pod-product-compliance
Lightning Source LLC
Chambersburg PA
CBHW040352220526
45473CB00009B/2865